# NUMBER 46

# HARDENING AND TEMPERING

FOURTH REVISED EDITION

## CONTENTS

| | |
|---|---|
| Modern Steel Hardening Plants - - - - - | 3 |
| Hardening Steel, by E. R. MARKHAM - - - - | 10 |
| Pack-Hardening Gages, by E. R. MARKHAM - - - | 15 |
| Forging, Hardening and Annealing High-Speed Steel, by W. J. TODD - - - - - - - | 19 |
| Local Hardening and Tempering, by WILLIAM A. PAINTER - - - - - - - - | 22 |
| Electric Hardening Furnaces - - - - - | 32 |
| Miscellaneous Hardening Methods and Suggestions - | 35 |

Copyright © 2013 Read Books Ltd.
This book is copyright and may not be
reproduced or copied in any way without
the express permission of the publisher in writing

British Library Cataloguing-in-Publication Data
A catalogue record for this book is available from the
British Library

# Blacksmithing

A blacksmith is a metalsmith who creates objects from wrought iron or steel. He or she will forge the metal using tools to hammer, bend, and cut. Blacksmiths produce objects such as gates, grilles, railings, light fixtures, furniture, sculpture, tools, agricultural implements, decorative and religious items, cooking utensils, and weapons. While there are many people who work with metal such as farriers, wheelwrights, and armorers, the blacksmith had a general knowledge of how to make and repair many things, from the most complex of weapons and armour to simple things like nails or lengths of chain.

The term 'blacksmith' comes from the activity of forging iron or the 'black' metal - so named due to the colour resulting from being heated red-hot (a key part of the blacksmithing process). This is the black 'fire scale', a layer of oxides that forms on the metal during heating. The term 'forging' means to shape metal by heating and hammering, and 'Smith' is generally thought to have derived either from the Proto-German 'smithaz' meaning 'skilled worker' or from the old English 'smite' (to hit). At any rate, a blacksmith is all of these things; a skilled worker who hits black metal!

Blacksmiths work by heating pieces of wrought iron or steel, until the metal becomes soft enough to be shaped with hand tools, such as a hammer, anvil and chisel. Heating is accomplished by the use of a forge fuelled by propane, natural gas, coal, charcoal, coke or oil. Some modern blacksmiths may also employ an oxyacetylene or similar blowtorch for more localized heating. Colour is incredibly important for indicating the temperature and workability of the metal: As iron is heated to increasing temperatures, it first glows red, then orange, yellow, and finally white. The ideal heat for most forging is the bright yellow-orange colour appropriately known as a 'forging heat'. Because they must be able to see the glowing colour of the metal, some blacksmiths work in dim, low-light conditions. Most however, work in well-lit conditions; the key is to have consistent lighting which is not too bright – not sunlight though, as this obscures the colours.

The techniques of smithing may be roughly divided into forging (sometimes called 'sculpting'), welding, and finishing. Forging is the process in which metal is shaped by hammering. 'Forging' generally relies on the iron being hammered into shape, whereas 'welding' involves the joining of the same, or similar kind of metal. Modern blacksmiths have a range of options to accomplish such welds, including forge welding (where the metals are heated to an intense yellow or white colour) or more modern methods such as arc welding (which uses a welding power supply to create an electric arc between an electrode and the base material to melt the metals at the welding point). Any foreign material in the weld, such as the oxides or 'scale' that typically form in the fire, can weaken it and potentially cause it to fail. Thus the mating surfaces to be joined must be kept clean. To this end a smith will make sure the fire is a reducing fire: a fire where at the heart there is a great deal of heat and very little oxygen. The smith will also carefully shape the mating faces so that as they are brought together foreign material is squeezed out as the metal is joined.

Depending on the intended use of the piece, a blacksmith may finish it in a number of ways. If the product is intended merely as a simple jig (a tool), it may only get the minimum treatment: a rap on the anvil to break off scale and a brushing with a wire brush. Alternatively, for greater precision, 'files' can be employed to bring a piece to final shape, remove burrs and sharp edges, and smooth the surface. Grinding stones, abrasive paper, and emery wheels can further shape, smooth and polish the surface. 'Heat treatments' are also often used to achieve the desired hardness for the metal. There are a range of treatments and finishes to inhibit oxidation of the metal and enhance or change the appearance of the piece. An experienced smith selects the finish based on the metal and intended use of the item. Such finishes include but are not limited to: paint, varnish, bluing, browning, oil and wax.

Prior to the industrial revolution, a 'village smithy' was a staple of every town. Factories and mass-production reduced the demand for blacksmith-made tools and hardware however. During the 1790s, Henry Maudslay (a British machine tool innovator) created the first screw-cutting lathe, a watershed event that signalled the start of blacksmiths being replaced by machinists in factories. As demand for their products declined, many more blacksmiths augmented their incomes by taking in work shoeing horses (Farriery). With the introduction of automobiles, the number of blacksmiths continued to decrease, with many former blacksmiths becoming the initial generation of automobile mechanics. The nadir of blacksmithing in the United States was reached during the 1960s, when most of the former blacksmiths had left the trade, and few if any new people were entering it. By this time, most of the working blacksmiths were those performing farrier work, so the term *blacksmith* was effectively co-opted by the farrier trade.

More recently, a renewed interest in blacksmithing has occurred as part of the trend in 'do-it-yourself' and 'self-sufficiency' that occurred during the 1970s. Currently there are many books, organizations and individuals working to help educate the public about blacksmithing, including local groups of smiths who have formed clubs, with some of those smiths demonstrating at historical sites and living history events. Some modern blacksmiths who produce decorative metalwork refer to themselves as artist-blacksmiths. In 1973, the Artist Blacksmiths' Association of North America was formed and by 2013 it had almost 4000 members. Likewise the British Artist Blacksmiths Association was created in 1978, and now has about 600 members. There is also an annual 'World Championship Blacksmiths'/Farrier Competition', held during the Calgary Stampede (Canada). Every year since 1979, the world's top blacksmiths compete, performing their craft in front of thousands of spectators to educate and entertain the public with their skills and abilities. We hope that the current reader enjoys this book, and is maybe encouraged to try, with the correct training, some blacksmithing of their own.

# CHAPTER I

## MODERN STEEL HARDENING PLANTS

From time immemorial when iron in its most crude form was introduced into the manufacturing and commercial field, it has been a well-known and accepted fact that heat with its varying degrees of intensity has a direct action on both the physical and chemical properties of the metal when the iron is submitted to its action; and, as a direct result, the entire structure of the iron is altered, and by altering or changing the methods of application of the heat treatment, any desired structure of the metal, either steel of cast iron, may be obtained. In spite of the fact that the truth of the above exposition was generally acknowledged, very little, if any, use was made of it; but as science developed, competition grew keener and keener, and the general cry in the manufacturing world became "reduced cost and greater output." To balance the effect of increased power and consequently larger machines, the working strength of the cutting tool, together with the working stress of the machine members, had to be greatly increased, and, during the past decade, the heat treatment has done more than its share in the work of accomplishing the desired results.

There are but few properly planned and equipped hardening plants. In the present chapter, however, two examples of first-class hardening rooms will be described, the one being that of the Worcester Polytechnic Institute, Worcester, Mass., and the other the hardening plant installed by Wheelock, Lovejoy & Co., in their New York store.

### The Worcester Polytechnic Institute Plant*

The Worcester Polytechnic plant consists of a room of spacious size in the design of which the comfort of the operator was well provided for. The temperature and ventilation of the room is controlled both by a fan and large windows which admit subdued natural light but exclude the direct sunlight, which is so undesirable in this kind of work. These windows are provided with shutters so that the natural light may be excluded; artificial illumination is obtained by means of incandescent electric bulbs. The room appears to a visitor, at first, somewhat like a dungeon, as the walls and ceiling are painted a "dead black," which color prevents any reflection of the various colored rays when the operator is experimenting on "color work." After this first impression has left the visitor and he has become accustomed to the light, the next thing that catches his eye is the row of various shaped furnaces placed symmetrically on the right side of the room. For convenience and simplicity, we will designate these furnaces (from right to left in Fig. 1) by the letters $A$, $B$, $C$ and $D$. Furnace $A$ (constructed

---

* MACHINERY, April, 1909.

4        *No. 40—HARDENING AND TEMPERING*

Fig. 1. Plant for the Heat Treatment of Steel in Worcester Polytechnic Institute

by the American Gas Furnace Co.) is built on the principle of the muffle furnace, is of the box type, and will readily heat a block of steel 8 x 4 x 14 inches. A temperature of from 2000 to 2100 degrees F. may be readily obtained by means of this heater, which is used to heat such work as requires an even heat and which would be destroyed by oxidation and the decarbonizing action of the air. Reamers, mandrels, taps and drills in their finished state are good examples of this type of work. Furnace B, known as the "barium chloride heater," is circular in form and lined with fire-brick, and the chloride solution is heated in a crucible built of fire-resisting material. This furnace is of sufficient size to accommodate all ordinary tools, and is employed to heat such grades of steel as require a rather high temperature, as high-speed steels, and which, at the same time, must be well protected in heating. This form of heat treatment is well adapted to those types and forms of tools which tend to heat unevenly, thus producing an unbalanced distribution of the shrinkage strains with the accompanying cracks. Furnace $C$ is of the same general design as furnace $B$, with the exception that this heater is made use of in connection with the lead bath. As the lead melts at a comparatively low temperature, this furnace is used when a lower temperature than that obtained with the chloride solution is desired, for example, when heating carbon alloy steel. Furnace $D$ is devoted to an entirely different operation, namely, oil tempering. Either linseed or machine oil is used in this heater, which is brought into action when the desired range of temperature is between the limits of 300 and 630 degrees F. The fuel used in all of these furnaces is the ordinary city gas, due to its convenience and ready accessibility, but oil fuel could be employed if so desired by the operator. As will be seen from the engraving, all the furnaces are provided with hoods of convenient form connected with an exhaust line, so that all poisonous fumes and gases from the lead, cyanide, barium chloride, etc., may be eliminated from the atmosphere of the room. At various and convenient positions about the plant are to be found rectangular tanks of convenient size, containing water and brine of varying densities. All the other baths, as for example, the various grades of oil and other cooling baths, are kept in covered cylindrical galvanized iron tanks. In order to properly care for and treat the air-hardening steels, an air jet is provided with a pressure of about 2 pounds.

The one feature which removes this plant from the class of the ordinary manufacturing establishment and places it in the ranks of those of scientific research and investigation, is its complete set of measuring instruments, including the Briston and Le Chatelier pyrometers and thermometers covering a range of temperature between the limits of 0 and 2960 degrees F. On one of the walls of the room is to be found the Bristol pyrometer, which is of the thermo-electric type, and consists of a permanent magnet moving coil type of galvanometer. The scale is graduated to read direct in degrees. Leads from the instrument extend over the entire room, so that it is a matter of a few seconds only to connect with the thermo-couple and obtain any desired temperature. If any question as to the accuracy of the instrument, or the action of gravity on its oscillating parts is advanced, a Le Chatelier pyrometer, operating on the same principle but having a vertical support, may be brought into action and the first readings verified.

As indicated by the above description, all grades of steel from the 15-point carbon steel to the high-speed, alloy, air- and water-hardening steel may be conveniently and efficiently handled and treated.

### Wheelock, Lovejoy & Co.'s Hardening Plant*

The illustrations, Figs. 2 and 3, show two views of a hardening plant installed by Wheelock, Lovejoy & Co., (selling agents for Firth Sterling

---
* MACHINERY, November, 1908.

Steel Co., McKeesport, Pa.) in the basement of their New York store at 23 Cliff St. The equipment is interesting in that it represents the latest development of gas furnace hardening and tempering baths. Fig. 2 shows a general view of the plant looking toward the street, while Fig. 3 is a view taken from the street end. The furnace in the rear, with a hood similar to the one in the foreground of Fig. 2, is for heating a chloride of barium bath, this being very successfully used for hardening "Blue Chip" steel, and the following description relates to the practice.

The tools to be hardened are first pre-heated, using the small American gas furnace shown next to the chloride of barium furnace. The pre-heating saves time in the barium bath, and is absolutely necessary to avoid checking or cracking the tools, as will be conceded when it is

Fig. 2. Hardening Plant, Wheelock, Lovejoy & Co., looking toward Street

known that the temperature of the barium bath is kept at between 2100 and 2200 degrees F. After the tools are pre-heated, they are immersed in the barium bath, being suspended by an iron wire, or, in the case of small parts, in sheet nickel baskets. The reason for using sheet nickel for the baskets is that chloride of barium has a slight dissolving effect on iron and the exposure of a large area of sheet iron in the bath would eventually destroy the baskets. Nickel is not affected to a perceptible extent, nor is the thin iron wire used to suspend ordinary tools.

The temperature of the barium bath is regulated by a Bristol thermo-electric pyrometer. This instrument, shown at the left in Fig. 4, is similar to a Weston ammeter or voltmeter, and the fire end is a thermo-electric couple. The heat of the bath effects the thermo-electric couple and generates a current that deflects the indicator of the indicating instrument to correspond with the temperature. For convenience in

## HARDENING PLANTS

operation, the indicating instrument is provided with a double hand, one hand, *A*, being controlled by the temperature of the bath, while the other, *B*, is a marker set by the operator to indicate the temperature which he desires to carry. This marker is made with a disk at the end that covers a hole in the indicating hand when the two coincide, as they do when the temperature has reached the predetermined point. Thus, an operator whose eyes are dazzled by the heat of the bath does not have to painfully study the graduations to see whether the pointer has reached the correct position, but by glancing at the instrument he can readily determine when the indicator is directly beneath the marker referred to.

The immersion of a piece pre-heated to a dull red immediately causes the indicator to drop, the temperature of the bath falling perhaps 30,

Fig. 3. View of Hardening Plant shown in Fig. 2, taken from Street End

40 or even 50 degrees. The fall in temperature is due to absorption of heat by the piece, being the same as the refrigerating effect of a lump of ice thrown into a pot of boiling water, and several minutes may be required to raise the temperature of a large piece to the temperature that is required. For hardening "Blue Chip" steel, a temperature of 2120 to 2140 degrees F. has been found most suitable. After this temperature is attained, the part is allowed to soak for a few moments, then is lifted out and dipped into the cooling bath shown at the right, Fig. 2, and left in Fig. 3, which consists of cotton-seed oil agitated by compressed air admitted at the bottom. The cotton-seed oil is contained in a large iron barrel surrounded by water in a wooden tub. The part hardened is allowed to remain in the bath until it is quite cold. In practice, the operator hardens a batch and then removes the pieces by means of a wire basket hanging immersed in the oil. It is recommended that milling cutters, end mills, slitting saws, etc., made of

"Blue Chip" steel, be used, in general, without drawing the temper. They will have the requisite hardness and toughness to stand up to the majority of work. However, an oil bath heated by gas and regulated by a thermometer is provided for tempering such tools as require it.

Chloride of barium is a white transparent salt ($BaCl_2OH_2$) which melts at a temperature of about 1700 degrees, the water of crystallization being driven off at a much lower heat. The salt volatilizes at an extremely high temperature, the loss at the temperature required for heating high-speed steel being negligible. The waste because of volatilization is, say, two pounds from a mass of barium weighing 75 pounds when held at a temperature between 2000 and 2300 degrees for five hours. This property of the chloride of barium bath of standing high temperatures without rapid volatilization is joined with others equally important. The piece heated is protected from the atmosphere during

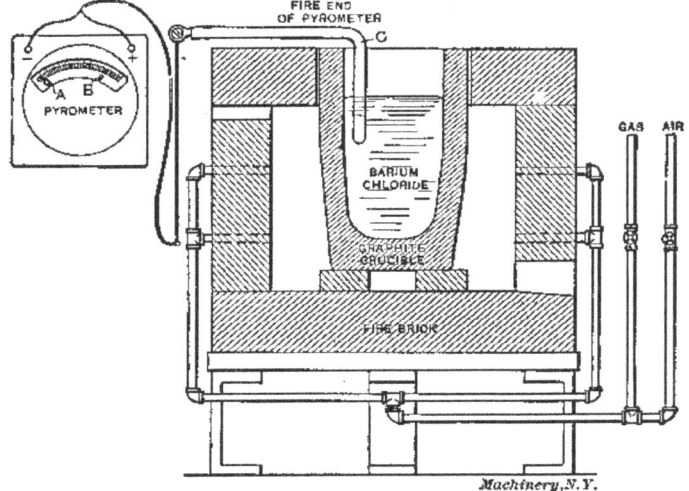

Fig. 4. Vertical Cross-section of Chloride of Barium Furnace

the heating period by the bath, of course, but the protective influence extends still further. A thin coating of barium clings to the piece when it is lifted out for immersion in the cooling bath, thus preventing oxidation. The effect of the barium on the steel seems to be limited to a slight mottling that quickly disappears under the action of cleaning and buffing wheels. The coating of barium remaining when dipped prevents the coating of burned oil so troublesome to remove, so that on the whole the process probably produces the cleanest work of any bath known.

Wheelock, Lovejoy & Co. have improved the furnace and crucible used for the chloride of barium bath. The common form of furnace and crucible in use employs a comparatively shallow crucible, which necessitates making a joint between the top of the crucible and the fire-brick cover. This gives trouble by loosening and permitting the hot gases to escape around the edge of the crucible. The improved construction illustrated in Fig. 4 utilizes a deeper crucible, the top of which comes flush with the fire-brick cover and simplifies the construction. The deep crucible also gives a greater volume of chloride of barium, consequently the refrigerating effect of the pre-heated steel parts, when immersed in the bath, is not so great. This illustration also shows the fire end, $C$, of the pyrometer immersed in the bath. It has been found advisable to employ crucibles made for steel melting, the ordinary graphite crucible used for brass melting giving trouble by flaking off into the barium.

The equipment of the plant includes an air compressor and exhauster, the former being required for the air blast in the furnaces and for agitating the oil bath, while the exhauster connected with the smoke pipes and hood, draws off the hot air and gases, thus keeping the working conditions fairly comfortable, even in the hottest weather. An efficient ventilating system is a prime requirement, inasmuch as the fumes of the barium are somewhat obnoxious and besides would have a serious rusting effect on the steel stock if permitted to pervade the basement where it is stored.

### The Use of Barium Chloride for Heating Steel for Hardening

In an article in the April, 1911, number of MACHINERY it was pointed out that barium chloride baths for hardening high-speed steel have certain disadvantages which prevent their usage in many cases. Tools heated for hardening in a crucible containing barium chloride have a soft scale or film of soft metal, perhaps about 0.003 to 0.006 inch deep all over the surface of the tool. Thus when heating high-speed steel to a temperature of from 2100 to 2400 degrees F., which is the proper heat for high-speed steel, the results are not satisfactory if the tools cannot be ground after hardening, so as to remove the soft scale or film of metal.

Tests made to ascertain the exact influence of barium chloride baths indicate that whenever this salt is used as a heating bath, it should not be permitted to reach a temperature of more than 2050 degrees F. It has been conclusively proved that high-speed steel tools heated in barium chloride do not stand as high cutting speed as do tools hardened by heating in an oven furnace. It seems that, particularly at high heats, some of the tungsten and carbon is removed from the tools into the bath, thus changing the structure of the surface of the tool being heated. When an amount of, say, 0.010 inch is ground off from the cutting edges of tools, the influence of the heating in barium chloride is less noticeable—in fact, sometimes not noticeable at all—but when the tools cannot be ground after hardening, barium chloride is not a heating medium which can be recommended under any circumstances. It is understood, of course, that for heating tools that can be ground after hardening, and also for heating carbon steel tools, the barium chloride bath possesses certain advantages.

# CHAPTER II

## HARDENING STEEL[*]

Every shop has one or more men who are considered authorities on hardening. In many cases the man is really an expert, is careful, and uses good judgment in heating the steel and in quenching in the bath; and if the piece is of sufficient size, he is sure to take the strains out by reheating directly after taking from the bath. In some cases, however, the success of one operation is measured by the failure of others. Thus if the steel passes through the fiery ordeal with enough of it left intact to do the work it is considered a *successful operation;* if not, the fault *must* be in the steel. A manufacturing concern once changed the brand of tool steel they were using three times in less than a year, because the man doing the hardening reported adversely on each make, after attempting to harden it. The brands furnished were from three of the leading makers of tool steel. After receiving repeated complaints in regard to the man's inability to harden the steel successfully, one of the makers advised the manufacturer to let some expert in hardening try the steel. Some milling machine cutters were made from each brand of the rejected steel and sent to the steel makers. They all came back hard enough, without cracks, proving that the trouble was not in the steel.

An expensive steel is not necessarily a satisfactory investment, and a "cheap" brand may be *very expensive.* It is necessary to understand just what is needed in a steel for a given purpose. Some makers have different grades of steel for different purposes—one for taps and similar tools, another for milling machine cutters, etc.—while others put out a steel that is very satisfactory for most purposes. Each has a good argument in favor of his particular method of manufacture. In some shops it is thought advisable to use a grade of steel adapted to each individual class of tool; while in other shops, where detail is not followed as closely, this would cause no end of confusion. That part of the subject must be left to the judgment of the individual shop. But the treatment of the steel in the fire and the bath, in order to be successful, must be along certain lines. The successful hardener is he who finds out what particular quality is needed in the piece he is to harden—whether extreme hardness, toughness, elasticity, or a combination of two of these qualities. Then he must know the method to use in order to produce the desired result. The shape of the piece, the nature of the steel, the use to be made of the article, must all be taken into consideration. He must also be governed somewhat by the kind of fire he is to use.

---

[*] MACHINERY, February, 1902.

## Heating the Steel

Some brands of steel will not stand, without injury, the range of heat that others will; some require more heat than others in order to harden at all. When hardening, no steel should be heated hotter than is necessary to produce the desired result. With some brands that give off their surface carbon very readily it is not advisable to heat them in an open fire, exposed to the action of the blast and outside air, as the products of combustion extract the carbon to such an extent that the surface will be soft even when the interior is extremely hard. While this might not materially affect a tool that is to be ground, it would spoil a tap, a formed cutter, or similar article, whose outside surface could not be removed. In hardening anything of this nature in an open fire, it should be placed in a piece of tube or some receptacle, so that the fire cannot come in contact with it while heating. There are a number of gas and gasoline hardening furnaces made which have a muffler to receive the work. The fire circulates around the muffler but does not come in contact with the steel. Very excellent results may be obtained when one of these furnaces is used. The front can be closed by means of a door, thus keeping all outside air away from the work. It will be found a great advantage if several large holes are drilled in the door, these being covered with isinglass, to enable the operator to see the work without opening the door.

Taking carbon from the steel is not the only injury done to a high grade of steel when heated in an ordinary blacksmith's forge by a careless operator. Most inexperienced men are apt to use a small fire, particularly if they find one already built. It may be mostly burned out, but the operator will not care to take the time to get fresh coal, and get the fire to the proper heat; so he puts on the blast and endeavors to heat the work. After a time the piece has all kinds of heats, ranging from a low red to a white heat. The operator thinks it *averages* well, and dips it in the bath. If it comes out in one piece he is fortunate.

Heating in a small fire is dangerous business, as the work not only comes in contact with the surrounding air, but with the cold air from the blast, which will cause minute surface cracks, making the steel look as though full of hairs. It will also fill the steel with "strains," causing ends of projections to crack and drop off in the bath.

If obliged to use the blacksmith's forge, use plenty of good charcoal. Make a large, high fire if the piece to be hardened is of any size; keep it up well from the blast inlet, using only blast enough to keep the fire lively, and bring the piece to the proper heat, burying it well in the fire to keep it from the air. The lowest heat that will give the desired result should be used. This varies in different makes of steel, and must also be varied somewhat according to size and shape of the work. The teeth of a milling machine cutter will harden at a lower heat than a solid piece of the same size made from the same bar. Most steelmakers in their instructions advise hardening at a low cherry red. To the average man this is a very uncertain degree; his cherries may be of a different hue from some other fellow's. Most of the leading brands of tool steel in small sizes give the best results, however, when hardened just after the black has disappeared from the center of the piece, provided it is heated slowly so as to get a uniform heat. In no case should steel be dipped when there is a trace of black in it.

The higher a piece of steel is heated—to a certain degree—the harder it will be; but if it is heated higher than to this degree the grain is opened, making it coarse and brittle, and it will be very liable to flake off under strain. For this reason, in the case of cutting tools, it is best to harden at as low a heat as possible. If the work gets too hot, yet not to a point where it is burned, it is always best to allow it to cool until the red has entirely disappeared, then reheat to the proper degree and harden, and the grain will be fine; but if allowed to cool to the proper hardening heat and dipped, it would be as coarse as if hardened at the high heat, and would also be very liable to crack.

## Annealing

In hardening, a great deal depends on the annealing. It is as necessary to understand how to anneal properly as it is to know how to harden correctly. As generally understood, the purpose of annealing is to soften the steel, which is all right, so far as the person is concerned who works it to shape, but its relation to hardening is another matter. It removes all strains in the steel, incident to rolling and hammering in the steel mill and forging in the blacksmith shop. Experience teaches the hardener that it is necessary to anneal any odd-shaped piece or one with a hole or impression in it, after it has been blocked out fairly well to shape, a hole somewhat smaller than the finished size being drilled in it, and all surface scale being removed. The most satisfactory method to pursue is to pack in an iron box with granulated charcoal, not allowing any of the pieces to come within one inch of the box at any point. This box should then be placed in the furnace and kept at a bright red heat for a length of time dependent on the size of the steel. Pieces one inch in diameter should be kept at a red heat for one hour after the box is heated through; larger pieces should be kept hot correspondingly longer, allowing the work to cool off as slowly as possible. An annealing heat should be higher than a heat for hardening the same piece. The proper heat for annealing, in order that all strains may be overcome, should be nearly as high as for forging the same piece; in other words, the work should be heated to a bright red and kept so long enough to overcome any strain or tension liable to manifest itself when the piece is hardened. Tool steel for annealing should never be packed in cast-iron chips or dust, as this extracts the carbon to such an extent that there will be trouble when hardening is attempted. Packing too near the walls of the annealing box will have the same effect to a less extent, but will be more troublesome, as the carbon will be extracted from the surfaces nearest the box, and not affected anywhere else, making the hardening very uneven.

If not situated so that this method can be used, very satisfactory results may be obtained by heating in a large charcoal fire to a uniform forging heat. Put two or three inches of ashes in the bottom of an iron box; on this place a piece of soft wood board, put the work on it, cover with another piece of board, and fill the box with ashes. The boards will char and smolder, keeping the work hot for a long time. Some blacksmiths use a box of cold ashes, while others use cold lime; either way is liable to chill the piece, making it harder than if allowed to cool in the air, and if either material is used it should be hot to get good results. Excellent results may be obtained by heating in a muffler oven, as a very uniform heat of any degree may thus be obtained. It can be run any length of time, but when a piece is heated through in this way it takes a long time to cool.

## Hardening Baths

Hardening a piece of steel is generally accomplished by heating to a low red, and plunging in some cooling bath. As so much depends on the bath, it is quite necessary to understand the effects of the use of the different kinds. The one most commonly used is clear cold water, though many use salt and water or brine. For hardening small articles that must be extremely hard, the following will be found very satisfactory: One pound citric acid crystals dissolved in one gallon of water. For very thin articles a bath of oil is necessary. For hardening springs, sperm oil is very satisfactory; when hardening cutting tools, raw linseed oil is excellent. There are hundreds of formulas for hardening compounds, some of which are excellent for certain classes of work. Some hardening solutions are poisonous, and are dangerous to have around; but for ordinary work the ones mentioned are sufficient.

Many successful hardeners use water that has been boiled, claiming better results from its use than from fresh water. Small odd-shaped pieces are not so liable to crack nor to harden unevenly when the water is slightly warm.

## Examples of Hardening

We will now consider a few pieces of work to be hardened by the open-fire method. If we have a muffler furnace, so much the better, as with this it is easier to get certain results; but with care very satisfactory work can be done when the blacksmith forge is used. If it is a small tap, reamer, counterbore, or similar article we are to harden, it is best to heat it in a tube, bring it to a low red, and plunge it in slightly warm water, or in the citric acid solution. If it is a hollow mill, with a hole running part way through it, we should dip it in the bath with the hole up, or the steam will keep the water from entering the hole, leaving the inside walls soft. The steam would also have a tendency to crack the piece; but with the hole up when dipping, by working the piece up and down well in the bath, the steam can escape, and the water can get at the work. Much trouble may be saved the hardener if attention is paid to the steam likely to be generated, and some way provided to prevent its keeping the water from the work. Brine does not steam as readily as clear water; neither do the different acid solutions used by many.

In hardening a milling machine cutter, it is best to have a large high fire, to bury the cutter well in the fire, and to use only blast enough to bring the work to the required heat, which should be uniform throughout. If the piece has not been annealed after drilling a hole through it, remove it from the fire when red hot, then allow it to cool off slowly until the red has entirely disappeared, when it can be again placed in the fire and slowly brought to the required heat; it is then plunged in a bath of tepid water or brine and worked around well until it stops "singing." At this point it should be removed and instantly plunged in an oil bath, and left there until it is cool, when the strain should be removed by holding it over the fire until it is warm enough to "snap" when touched with the moistened finger. It can then be laid aside, and the temper drawn at leisure. In hardening punch-press dies we can treat them the same; if there are any screw holes for stripper or guide screws they should be plugged with fire clay or graphite.

Metal-slitting saws can be hardened nicely between iron plates whose surfaces are kept oiled. The saws should be heated in such a manner that the fire does not come in contact with them. It is best to heat on a flat plate, as the tendency to warp is much less than if laid on an uneven surface. When the saw is properly heated, place it on the lower oiled plate, placing the other one on it as quickly as possible; hold the upper plate down hard until the saw is cool. If there are many such pieces to harden, a fixture can be made so that one man can handle the saws and fixture alone—otherwise it requires two operators.

If there are no other means of drawing temper, the work may be brightened and drawn by color; but, if possible, do the drawing to temper in a kettle or crucible of oil over the fire, gaging the heat by a thermometer. Much more satisfactory results can be obtained by this latter method; and if very many pieces are to be done, it will be found much cheaper. A very light yellow is 430 degrees; a straw color is 460 degrees; a brown yellow, 500 degrees; a light purple, 530 degrees. A milling machine cutter for ordinary work should be drawn to 430 degrees; a punch-press die to 500 degrees; the punch to 530 degrees, and metal-slitting saws to 530 degrees.

# CHAPTER III

## PACK-HARDENING GAGES*

Pack-hardening, as the term is generally applied, consists in treating steel, generally tool steel, with some carbonaceous material until it will harden in oil. It is well known that steel hardened in oil is less liable to spring than when hardened in water. The tendency to crack is almost entirely done away with, unless the steel is improperly treated in the fire, and has the maximum of toughness. Now, if we are able to treat the steel so that it will be as hard as though dipped in water, and yet have the toughness due to oil-hardening, and at the same time reduce the tendency to spring to the minimum, it would seem that we have the ideal method of hardening.

The process consists essentially in supplying the surface of the steel with an additional amount of carbon by some material that will not in any way injure the steel. In order to provide the additional carbon, the steel must be packed in iron hardening boxes with the carbonizing material. Some have used charcoal for this purpose. While charcoal is a carbonizing agent, and is used many times in case-hardening machinery steel, and also in the cementation process for converting wrought iron into steel, yet its effect on high-grade steel in the process of carbonizing is not satisfactory, as it renders the steel coarse, and very similar to blister steel. No form of bone should be used when pack-hardening tool steel, as bone contains a high percentage of phosphorus, and the effect of this is to make steel weak and brittle.

For steel that does not contain more than 1¼ per cent carbon (125 points), charred leather gives the best results. Above this percentage use charred hoofs, or horns, or a mixture of the two. The leather, hoofs, or horns, may be used over and over by adding a quantity of new material each time.

The work should be packed in the hardening boxes so that no part of any piece of work comes in contact with the boxes; in fact, there should be at least ½ inch space between the work and the box. A layer of the carbonizing material should be placed in the bottom of the box, and a layer of work placed on this, taking care that no two pieces touch each other. If we are treating gages, or pieces of steel that are apt to spring unless care is used, we should make sure that they are so placed in the box that there will be as little liability of springing as possible when they are drawn up through the packing material. They must not be dumped into the hardening bath, as is the case when ordinary case-hardening is done.

In order to be able to properly handle the work, each piece should be wired with a piece of *iron* binding wire, as shown in Fig. 5, and

---

* MACHINERY, June, 1908.

the pieces so placed in the box that there will be the least resistance possible when drawing them out. At times they may stand on edge, as shown in Fig. 5. For certain shapes, however, it is advisable to stand them on end.

When several layers of work are packed in a box, the wires should be so arranged around the edge of the box that the various layers may be taken out in order, commencing with the top row. This is easily accomplished by marking the sides of the box with chalk, designating the side where the top row of wires is, as 1, the one where the second row is, as 2, and so on. Unless we adopt some such method, the pieces get all mixed up, and some will be drawn to the surface of the packing material, and will cool before the operator has a chance to dip them.

As in all heat treatment of tool steel, the heats should be as low as is consistent with desired results, and the heats must be uniform

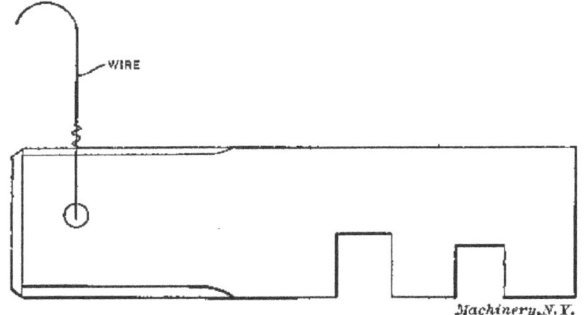

Fig. 5. Piece to be Hardened and Wire for Handling

throughout the box. It is also necessary that we gage the length of time the steel is exposed to the action of the carbonaceous material. Unsatisfactory results follow any attempt to gage the length of time by the time the boxes are in the furnace. In order that the operator may know when the contents of the box are heated, holes are drilled through the cover of the box at the center, and test wires are run down to the bottom of the box, as shown in Fig. 6. These wires should project about one inch above the top of the cover. The holes in the cover may be of any size to accommodate the wire to be used; a good size is 1/4-inch holes and 3/16-inch wire. When the box has been in the fire, according to the judgment of the operator, until the contents are heated to a low red, a wire may be drawn, by means of long tongs, and its condition noted; if it is red hot, begin timing the heat; if it is not red, wait a little while, and drawn another. Continue doing this until one is drawn that is of the desired temperature. The wires passing down at the center of the box, and between the pieces, will not be red until the pieces are of the same temperature.

The length of time necessary to expose the pieces to the action of the heat depends upon how deep we wish to harden the steel. For ordinary snap gages 1½ to 2 hours after the steel is red-hot is sufficient, but the time must be varied according to the percentage of carbon that the steel contains and its intended use.

# PACK-HARDENING GAGES

Sometimes locating gages are made with the gaging holes made to the finished size of the gage. This method is not to be advocated where it is possible to use hardened bushings. In the latter case, the holes in the gage may be made of the proper size for the bushings, and the gage left soft, while the bushings are hardened, ground and lapped, and pressed into place without any tendency to distort the gage. But when it is necessary to make the gage of one piece, and have the gaging holes to size in the gage without bushings, then the pack-hardening

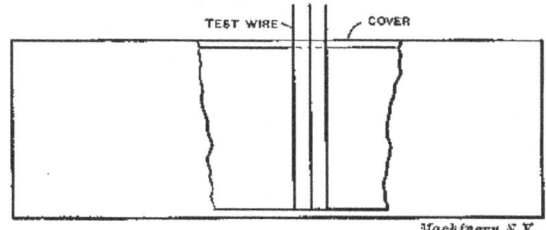

Fig. 6. Hardening Box with Test Wires which Enable the Operator to Determine when to Begin Timing the Heat

method will be found to work satisfactorily, as the heats may be very low, and the tendency to distortion will be eliminated, provided the processes of annealing and machining have been properly done.

As an example of pack-hardening gages, a case from practice may be cited. The gage was of the form shown in Fig. 7, and it was necessary that the walls of the opening through the gage be hard, yet the opening must retain its shape. The hole was filled with finely pulverized charred leather; the handle and the portion connecting it with the body were covered with fire clay which was wound with fine iron bind-

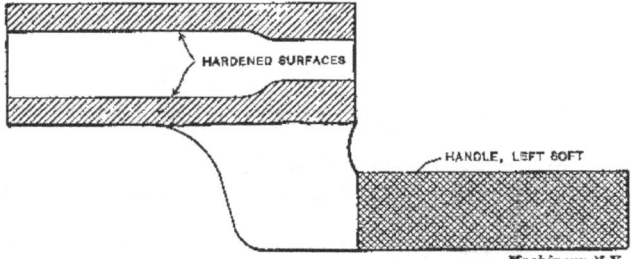

Fig. 7. A Gage to be Hardened as Indicated

ing wire to prevent it falling away when baked. The gages were packed in scale collected in the forge shop. This scale came from the outside of pieces of iron and steel as they were being forged. Being free from carbon, they absorbed or took the carbon from the surface of the steel. The ends of the opening through the gages were covered with fire clay mixed with water to the consistency of dough, which was allowed to harden before the gages were packed. The fire clay prevented the carbon gas escaping from the leather, as the scale would have taken it up very quickly.

When the gages had been exposed to the carbonizing influence of the leather for one and one-half hour, they were removed from the box in which they were heated, placed in a bath of raw linseed oil, one at a time, and a jet of oil was pumped through the opening after the leather had been removed. The fire clay around the handle and the portion connecting it with the body was left on until the gage was hardened, when it was removed. The walls of the hole were found to be hard, and as the surface of the gage was practically decarbonized, there was little danger of its pulling the piece of steel out of shape. The handle and adjoining portion, being protected by the fire clay, did not harden, or even cool quickly enough to distort the gaging portion in any way.

Many times receiving gages are made of several pieces which are fastened to a plate as shown in Fig. 8, which is a receiving gage for a

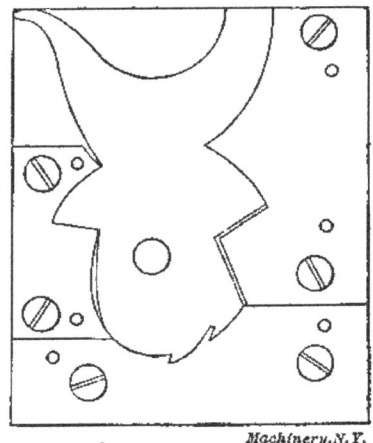

Fig. 8. Receiving Gage of a Type Advantageously Hardened by the Pack-hardening Process

gun hammer, and it is necessary that the various portions be gaged accurately, and that each portion bear a certain relation to every other portion.

As shown, the various portions of the gage are made in sections, fitted in place and hardened. Unless these pieces are hardened by some method that eliminates the tendency to spring, they will be of little use after they are hardened. This is a case for pack-hardening. Pack the pieces in leather in a small iron box, run for one hour after they reach a low, red heat, and harden in raw linseed or sperm oil. It will not be found necessary to heat the steel treated in this way as hot as if heated in an open fire and dipped in water. It is not necessary to heat steel in the form of gages quite as hot as if it were made into cutting tools; but, even in the latter case, be sure to keep the heat down, and do not dip in extremely cold oil; have it warm, but not hot.

# CHAPTER IV

## FORGING, HARDENING AND ANNEALING HIGH-SPEED STEEL[*]

The rapid progress of tool steel manufacture, and particularly the advent of high-speed steel, makes it more and more necessary for the successful toolsmith to devote some of his spare time to a careful study of the nature and adaptability of new steels with special reference to their use in tool-making. The ease with which high-speed steel may be forged, shaped and hardened permits its use in the construction of many tools and dies. Hence, we not only find it in use in the machine shop for cutting tools, but in the press room for the construction of forming dies for hollow ware; also for wire-forming dies, and in the bolt mill for bolt-heading dies on hot work. It is used for shearing dies for hot and cold bar work and for many other purposes unthought of in connection with air-hardening steel. But, unfortunately, owing to the number of different makes of alloy steel now in use, no standard rule can be adopted for manipulation. Therefore, in order to determine the proper forging and hardening heats, it must be a matter of experiment for the toolsmith, and he will find it a very good method to test the particular brand he is about to use by giving a piece of it as much heat as he thinks the steel can properly stand and noting the result.

### Forging High-speed Steel

The majority of the high-speed alloy steels should be heated, whenever possible, in block fire with a good body of forge coke. Thoroughly heat the steel in a clean, bright fire to a full yellow plastic heat, which means a temperature of 1650 to 1850 degrees F. This, of course, is a much higher temperature than can safely be used in forging air-hardening or carbon steel. In fact, the heat treatment given alloy steels of the high-speed variety is so diametrically opposite the generally accepted practice of working carbon steel that it seems to be a difficult matter to persuade the average smith to raise the heat to the proper temperature when working high-speed steels, but when this fact is once impressed, he will be surprised at the ease with which it can be worked. By rapid forging and repeated and thorough heating, it will be found no difficult matter to give alloy steel almost any desired shape, but, unlike carbon steels, it will not do to run the heat down low in forging, as in such practice the steel will most likely be found cracked or split. To heat thoroughly and often is the essential point to remember.

---

[*] MACHINERY, February, 1906.

## Annealing High-speed Steel

After forging alloy steel for lathe, planer and shaper tools, it is, of course, unnecessary to anneal them except to let them cool down in the air in a dry place. When cold, rough grind to a proper edge; then they are ready for hardening, which process will be described later. In making use of alloy steel for taps, reamers, twist drills, milling cutters, or forming dies, it will be necessary to anneal the forgings in order to enable them to be worked in the machine shop.

The different sizes and shapes of this steel can generally be procured already annealed from the steel-maker, but when unable to obtain the annealed bar, and the work is required to be annealed, proceed in the following manner: A very successful method of annealing high-speed steel, and also self-hardening steel, consists of placing the tool in a wrought-iron or cast-iron tube, having space large enough in circumference and length to accommodate the work and plenty of packing material. One end of the tube can be threaded for a cast-iron cap, which can be screwed on and off as desired; the other end can be permanently fixed on the tube, as it is generally only necessary to use the one end. Have a number of ⅛-inch holes drilled in the tube, and also a few 3/16-inch holes in the end caps. The idea of the holes in the tube is to procure a vent for letting off the gas which is generated when heating the packing material. The end holes can be used for a number of test wires, which can be withdrawn as the heating progresses, and by this means the operator will be able to ascertain the proper heat desired, which should be a bright orange or a trifle higher, according to the nature of the steel. When the desired heat is reached, regulate the blast sufficiently to hold the muffle and its contents at this heat for a period long enough to allow the heat to thoroughly penetrate the steel. Then the muffle with its contents may be buried in dry slaked lime or sawdust and ashes, and allowed to cool down slowly. If a furnace is used for the heating, allow the muffle to stay in until furnace and all cools down, when the steel will be found quite easy to work in the machine. Respecting this "dead heating" in the furnace, unless the steel is properly packed in the muffle in order to exclude the oxygen blown into the furnace by the blast, the steel is apt to oxidize and the carbon content thereby becomes lowered, resulting in overannealed steel. The packing material used may be any one of several kinds now commonly used for other work, but charred leather gives the best results, and dry, fine smith coal of a good, clean quality is effective.

## Hardening High-speed Steel

The hardening of high-speed steel can be done in numerous ways according to the requirements desired for the tool. For turning and cutting tools for lathe, planer, shaper and slotter, it will be found a good practice to heat the nose of the tool slowly to a bright red; bring it rapidly with a good, quick fire to a white fusing heat, about 2200 degrees F., on the point, then quench it quickly in oil and leave until it is cold. It must be borne in mind, however, that when a tool is at this high fusing heat it must not come in contact with the fuel of the fire, as, should that occur, the nose of the tool would be ruined and require considerable grinding or reforging into shape. After regrinding the tool on a wet stone it will be ready for work and should give a good account of itself, and incidentally reflect credit on the smith who worked it.

The oil used may depend on how hard the tool is desired. We will suppose it is required "dead" hard on the cutting edges; this is as hard as it is possible to make it for machine shop use and still retain sufficient toughness. To obtain this result, after fusing the point or cutting edges, quench quickly in thin lard oil, or for extreme hardness, quench the tool in kerosene oil, when about the maximum hardness of this steel may be obtained. In using oils, especially the kerosene, great care should be taken, or the oil may flame and burn the operator. The oil tanks used for hardening should be constructed preferably of galvanized iron, fitted with close-fitting covers, and provided with a screen a few inches from the bottom on which to rest the work in order to facilitate the quenching by allowing free circulation of the quenching bath to all parts of the tool.

In hardening machine-finished work, such as dies, milling cutters, taps, etc., made from this steel, we must proceed differently, as the course described would not answer for all purposes. All machine-finished work is to be packed in muffles or iron boxes as described for annealing, bringing the heat up in a similar manner, and testing as described. When heated sufficiently, remove the cover of the muffle and quench the tool in a bath of oil composed of half parts each of thin lard and raw linseed oils, keeping the tool in motion while in the bath until it has cooled to a little below the boiling point of the oil. Then it can be removed from the bath and left to cool in a warm, dry place, when, after being cleaned and polished, the temper may be drawn to a dark straw for general purposes. 1 may add that it is not always necessary to draw the temper on this character of tools; neither is it always necessary to harden such tools as milling cutters, dies, reamers, taps and drills in the oil. Very often it will be found only necessary to heat them to a good, bright red heat about 1500 degrees F. and lay them in a cold blast; when cold they will be hard enough for ordinary work. The objectionable feature of the cold-air blast for such work is the liability of the steel to scale on the finished parts, possibly causing a slight loss in size, though this seldom occurs when proper care is used, and the steel is not heated to a scaling point. The cold-air treatment, however, is applicable to work such as forming dies, wire dies and tools required to withstand some shock.

Sometimes it is unnecessary to pack the work, as it may be successfully heated by just using the tube and closing one end only. Place the tube in the fire and the work in it, bring it to the required heat with a gentle blast, and turn the tube frequently to insure an even heat, and to prevent it from burning.

# CHAPTER V

## LOCAL HARDENING AND TEMPERING*

In describing the shield process or method of local hardening and tempering of tool steel, some of the results of over four years of experimenting and practical application are recorded. The process is a radical departure from any practice in use up to the present time and acts on a principle that appears not to have been recognized heretofore. The wearing qualities of dies and tools so treated have been demonstrated by every-day use, and the efficiency of these as compared with tools treated by the conventional processes is decidedly superior. Breakage is reduced to a minimum, both in hardening and tempering and in use.

In using the shield method, the same judgment and common sense must be employed as in hardening by the regular method. If the hardener is hardening a piece that should be dipped in clear water, following the usual practice, it should be dipped in clear water when provided with the shield, and allowed to cool with the shield on. Or, if the piece is complicated and of a delicate structure that requires cooling in oil, the same practice should be followed with the covered piece, dipping it in oil in the same manner. Again, if a special bath is required for certain work, use the bath in the same way for the shielded work. The results are exactly the same, except that with the shield method only the parts exposed harden, while the parts covered remain soft and in their natural condition.

### Heating

For heating tool steel to be hardened by the shield method, the operator has the choice of any fire he prefers. An open forge answers the purpose for small work, but the furnace is better on account of the more even heat. A lead bath cannot be used because of the lead getting between the covers and the steel to be hardened, which will cause an explosion when dipped into water. This is not a serious drawback, as the lead bath has no marked advantages over a good coke or gas furnace. Moreover, the lead bath has a disadvantage in that it leaves a thin film or coating on the steel, which tends to keep the cooling bath from penetrating to the steel and cooling the piece as rapidly as it would cool if heated in a furnace.

### General Application of the Shield or Cover Method

The shield method can be applied to almost all tool steel. Any brand or grade that is desirable to use can be handled in exactly the same way as without the shield. The advantage of the shield is that the essential wearing surfaces are hardened, and if these parts only

---
* MACHINERY, August, 1908.

# LOCAL HARDENING AND TEMPERING

require hardening, why harden the piece all over? A piece hardened all over is weakened because of the internal strains and stresses which are likely to develop cracks and fractures and perhaps ruin an expensive part after it has been made all ready for use.

Gages of all kinds for external and internal use can be treated to advantage, and many gages now being made of machine steel and case-hardened can be made advantageously of tool steel in this manner, only the wearing parts being actually hardened, while the remainder is left soft. Multiple-throw crank-shafts for automobiles and high-duty automobile parts are readily treated by the shield method so that the actual wearing parts are hardened while the remainder of the piece is unchanged. Tool steel bearings for machinery may be hardened on the inside only, whether the bearing is in one piece or is split in two parts. Large rolls for rolling mills can be made with the journals soft or partly soft, thereby avoiding their breakage at the shoulders. Hammer faces for pneumatic hammers can be treated at the shank so that the shank is soft while the die part is hard, thus avoiding the breakages now generally experienced. All kinds of drill jig bushings can be hardened, either externally or internally, as desired. Shear blades and other large pieces for similar work can be hardened on the cutting edges while the bulk of the metal is left soft.

## Materials Used for Shields

In practice the thinnest possible cover to accomplish the desired results should be used. Material thicker than 0.020 inch for ordinary work is not required, but not thinner than 0.010 inch for small work should be used. The thickness that serves best for average work is about 0.014 inch. The object of the shield is to cool the steel as quickly as possible under the cover without hardening it and without leaving a line of tension between the hardened and unhardened areas. The cover or shield can be made of any sheet iron or sheet steel; scrap pieces, even if rusty, answer the purpose satisfactorily. Galvanized iron or tin plate should not be used, as the coating of zinc or tin when heated enters the steel and deteriorates it.

The shields or covers are made in different ways to suit the conditions. Some are made with top and bottom pieces with another strip or piece for the edge to act as a binder to hold the top and bottom in place and to protect the edges of the piece. Some shields are made in one piece where the shape is not irregular. This applies in the case of round drawing dies and similar parts. In such cases, of course, the shield is formed to shape by cutting and bending. Where an irregular shape is to be cut out, the piece to be hardened is laid on the sheet metal and the outline scratched on it. The shape is cut out with a chisel, leaving a margin of the required width around the edges that are to be hardened. The width of this margin depends on the size of the part and the nature of the work. A pair of hand shears or snips, a small chisel, a scriber, hammer, punch and pliers are the principal tools necessary. A supply of sheet metal, rivets, and iron binding wire completes the outfit.

## How the Shield Method Differs from Other Protective Methods

Mechanical journals and books on shop practice have not hitherto treated of this method, although other methods very similar have been described. There are several methods moderately successful for local hardening and tempering, one being by dipping the piece part way into a bath, a practice that is commonly followed in hardening chipping chisels. Another method is holding a heated piece over and above the bath and directing a stream of water onto the part to be hardened. Large, round drawing dies having to be hardened on the inside are examples of this practice, a cone-shape spout being used at the bottom of the pipe to deflect the stream into the interior of the die. A third method is that of making a box or form of heavy steel, cutting or drilling out the shape of the piece to be hardened. A lid is fitted to the box and the piece is heated in the fire, and after being heated is placed in the box and quenched. This method is objectionable in that the box, being made of thick material, causes uneven

Fig. 9. Blanking and Threading Dies in Protective Covers

cooling, most of the heat having to be drawn off through the exposed parts of the piece. A fourth method is the use of fire-clay or asbestos wrapped around the parts to be left soft, the holes being plugged with fire-clay, asbestos or even putty. Iron wire is used to wrap around a non-conducting material to hold it in place. It then has to be baked in an oven or furnace until the moisture is dried out. Then the part is heated in the furnace to the hardening heat, and dipped. Inasmuch as these materials are poor conductors of heat and as the heat must be drawn quickly in order for the steel to contract evenly with the hardened or exposed parts, this method has not been very successful.

A piece of steel heated to a proper hardening heat and quenched in a box or shield of very thin steel or iron material, will remain soft because the water cannot cool it directly. Steam or vapor forms between the cover and the steel, driving the water away until the piece has cooled down. If holes are cut through the cover to let the steam out more rapidly, that part of the steel exposed will harden. To show that it is the steam imprisoned between the steel and the shield which is sufficiently non-conductive to produce the difference between hardening and annealing, take a round piece of tool steel and fit a cover tightly around it, leaving half of its length exposed. Heat this piece all over and cool all over. It will be found that the steel has hardened its full length under the cover, as well as on the exposed parts. The reason is that when the cover is tightly fitted, no water can penetrate between the shield and the steel, and hence no steam is formed. The cover being thin, the steel cools rapidly and hardening takes place. This proves that the cover itself is a good conductor of heat, and that it is the thin stratum of steam which will always exist between the cover and the piece as ordinarily fitted, which is sufficient to check the hardening process.

### Examples of Work Locally Hardened

The part shown in Fig. 9 at A is on ordinary blanking or cutting die and is an example showing how any shape to be hardened can be outlined in the cover. The cover is made in two pieces, the top and sides being one piece and the bottom the other piece. The material of this shield is No. 26 sheet steel and is about 0.018 inch thick.

The second piece shown in Fig. 9 at B is a round threading die, and the shield is applied so that the threads only are hardened, the ex-

Fig. 10. Square Threading Die and Press Dies with Cover in Place, Ready for Hardening

terior remaining soft. It illustrates a very practical use of the shield method. The die is split at one point, being of the adjustable type. Inasmuch as only the teeth are hardened, there is no danger of cracking the die in adjustment, and there is less distortion and change of pitch. The cover is made in one piece and can be used several times before being destroyed by the heat.

The third piece shown in Fig. 9 at C is a plain blanking die, but instead of being hardened around the edge of the cutting part as at A, it is hardened half the thickness of the steel, leaving the whole lower half soft, this being protected by the cover, while the upper half is hardened. This gives the die all the advantages of a die made of composite steel, a material that is made of half tool steel and half soft steel or iron welded together. This material is extensively used on small die work, but the shield method makes its use unnecessary as all its advantages are obtained very simply, without the risk of splitting at the weld, as sometimes happens with composite steel.

A square threading die is shown in the shield at A in Fig. 10, and the same remarks apply in regard to distortion and change of pitch as in the case of the round threading die. At B is shown a press die in its cover ready to be hardened. The exposed portion is not large in proportion to the size of the die, but it will harden just as well as if there were a half-inch margin exposed all around the edge.

The cover is held on by iron wire. Brass wire or rivets should not be used, as the heat destroys them. At *C* is shown a progressive or following die. There is nothing of special interest in the design, but the method of hardening a narrow ring around each hole is unique. The cover is cut out so as to leave a narrow margin of exposed steel around each hole. The die is hardened exactly as the cover shows, the remainder of the body being as soft as before dipping.

Fig. 11, at *A*, shows a forming die for a drop-hammer. It is 8½ inches long, ¾ inch thick and 4 inches high, and has been hardened all around the outline of the top or face for a width of ½ inch. The scale is still on the steel and no finishing has been done. The cover had two large holes cut out to balance the cooling as the narrow center would warp the bottom. This piece was hardened without warping. At *B* is shown a large blanking punch with the face up. In hardening large punches it is only necessary to harden an area around

A        B
Fig. 11. Drop Hammer Die and Punch with Shields

the cutting edge. The steel plate or cover on the face of the punch is 0.020 inch thick and was held to the face of the punch by three 3/16 inch screws, holes being drilled in the face of the punch to correspond. This is a good example of cover hardening and demonstrates its value, as the punch retains its shape perfectly after being dipped.

Two drawing dies are shown in Fig. 12 with the covers on ready to harden, and a third die is shown without a shield. These shields are made in one piece and can be used over again. A drawing die needs to be made as hard as possible, and yet when a die is hardened all over it is necessary to draw the temper in order to relieve the internal strains. By the shield method the working part is made as hard as it can be made while the soft part gives it a suitable backing and strength. The shrinkage can be regulated by the thickness of the shield or by putting vent holes around its edges. The vent holes let out the steam and cause the shielded part to cool more rapidly. Dies hardened by the shield process can be shrunk repeatedly to take up the wear.

A reverse drawing die 6 inches outside diameter, 5 inches inside diameter, with a hardened ring riveted to the top or wearing edge, is shown in Fig. 13 at *A*. At *B* is illustrated the method of hardening the ring, leaving the shoulder or bottom edge soft by covering it with a shield of thin steel. The ring can be fitted after hardening, driven

## LOCAL HARDENING AND TEMPERING

on a shoe and holes drilled through from outside of the machine steel shoe, and riveted. The hardened portion is then ground to size. If the top rings were welded to the shoe it would have to be discarded when worn down.

In Fig. 14 is shown an embossing die made of tool steel. It is 8½ inches in diameter and 3 inches thick. This die was covered on the bottom and sides with No. 25 sheet steel, 0.020 inch thick, and placed in a basket or hanger so that it could be immersed by a tackle. The

Fig. 12. Two Drawing Dies Prepared for Local Hardening and One Without Cover

embossed part was filled with bone dust for protecting it in the fire, but charcoal or any suitable filling will do as well. When covering shallow dies, the top edge should be left standing about half an inch above the face of the die instead of being folded over, as shown in the illustration. This face is filled in as described above, as the filling will protect the face of the die. It should be left in, when dipped, and the die should be suspended face up and about 3 inches below the water level, while a stream of water of large volume but low pressure is directed against the center of the die while the die is

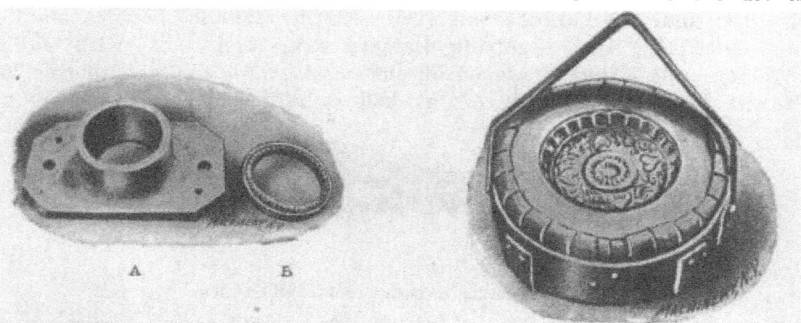

Fig. 13. Reverse Drawing Die        Fig. 14. Embossing Die

moved around in the tank. This can be done with one hand holding the tackle, while the other hand directs the stream. The filling or packing will float away, leaving the surface of the die clean. A ¾-inch hose without nozzle is about right for ordinary work.

In this connection it may be mentioned that a skeleton platform with hanger is an advantage for handling large work, as any piece too large for tongs can be placed on the platform and lowered into the tank. It may be cooled from the bottom, if necessary, as well as from the top. Some shapes give better results if covered only on the sides, leaving the top and bottom exposed. The hanger gives the bottom a chance to cool off, but not as rapidly as the top on which the stream of water is directed. This method is an imporevement over the present way of hardening drop forging dies, this class of hardening being regarded as very difficult where the shapes are irregular and the thickness of the walls uneven. It will apply to all deep or hollow dies.

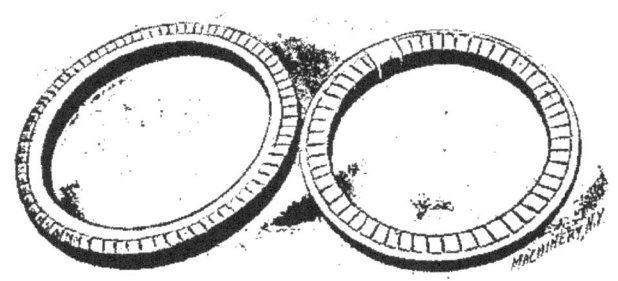

Fig. 16. Blanking Die and Punch

To the left in Fig. 15 is shown a blanking die 11⅛ inches inside diameter, 1¼ inch thick and 1¼ inch width of walls. This die is covered and hardened in the regular way. The temper is drawn, after which it can be placed in the lathe and trued up. The outside is as soft as it was before dipping. Obviously, this die is much stronger than one hardened all over. At the right is shown a blanking punch 11⅛ inches outside diameter, 1⅛ inch thick, with walls 1⅛ inch wide. It is hardened on the outside only and the temper is drawn. The inside is as soft as before hardening, and is made a

A       B       C
Fig. 16. Samples of Lathe Tools with Shields

driving fit on a cast-iron holder. The fit can be made a very tight drive without danger of cracking the ring.

Fig. 16 shows the application of the shield method to hardening lathe tools. At A is shown a right-hand side tool of carbon steel hardened on the cutting face only, the back being left soft. The cutting edge can be left much harder than usual without danger of weakening the tool. At B is shown a self-hardening tool that has been heated up to the proper temperature without cover, and the cover applied to the tool before dipping in oil. This method works equally well on air-hardened tools, the cover being put on before holding the tool

In the air-blast. The breakage of self-hardening steel tools is an expensive item and usually amounts to more than the actual wear of the tools, but if they are made with a soft back or partly soft back and

Fig. 17. Miscellaneous Examples

with only the working parts hard, as will be the case with the shield process, the tool is much stronger, and is much less liable to fracture in service. At *C* is a half diamond point lathe tool which is hardened only on the working face. The sheet covers can be applied to

Fig. 18. Group of Large Dies and Covers

any shape of lathe or planer tool by any one of ordinary mechanical skill.

In Fig. 17 a ball-peen hammer is shown made ready for hardening; the eye for the handle is not filled, but is covered with a shield of steel wrapped around the part to be kept soft. The eye will not harden, as the heat of the steel inside of the cover, when dipped, creates steam which drives the water back until the heat is reduced. The milling cutter in this illustration is shielded with two circular plates 0.020 inch thick, and the hole in the center is not filled. The disks of sheet metal are held together by iron wire, and any desired area

Fig. 19. Another Group of Large Dies

on the sides of the cover can be left soft by making the size of the disks to suit. This is far better than the heavy plates that are generally used, as the heat is drawn off more rapidly, leaving less internal stresses in the steel. The tap shown has been hardened on the teeth only, the body and shank being shielded so that they are left soft.

Fig. 20. Miscellaneous Collection of Shields or Covers Showing Range of Application

Taps hardened in this manner are truer in the pitch and the size is less likely to be changed than if hardened all over. The preparation of the cover took less than 5 minutes' time, only two strips of sheet steel and a short section of iron binding wire being required. Flat drills are hardened on the point and sides only, the center being left soft. Such drills will not snap off short when caught in the work, but will bend before breaking. The shield consists simply of two thin plates held on by iron wire.

Fig. 18 illustrates a group of large dies and their covers. In Fig. 19 is illustrated another group of large dies, and in Fig. 20 is shown a miscellaneous collection of covers that have been used to shield work from ¼ inch diameter to 12 inches diameter. The shields are made mostly from stock 0.010 to 0.020 inch thick. Some of the circular shields have been in the fire several times.

The shield method of local hardening is applicable to a vast range of work, and the cutting tools, dies, taps, gages, etc., locally hardened and tempered, are less likely to warp and change in shape in hardening and are more durable in service than when hardened in the common way. The application of the shields is cheap, and any one of ordinary skill is able to cut them out and apply them to the work. The saving in breakage alone will more than pay the expenses of applying the shields, while the greater durability of tools so hardened makes the process profitable to use.

# CHAPTER VI

## ELECTRIC HARDENING FURNACES*

In externally-fired furnaces, the heat losses are always considerable, and only a small part of the energy used in heating is utilized for raising the temperature of the metal to be hardened. There is also a disadvantage in employing gas or oil-fired furnaces in that the high temperatures rapidly destroy the crucibles. Electric hardening furnaces, therefore, possess marked advantages for this work over the various types of externally-fired furnaces. The electric furnace described in the following has been brought out by the Allgemeine Elektricitäts-Gesellschaft of Berlin, Germany. A bath of melted metallic salts is contained within a fire-brick crucible, inside of which, at two opposite sides, are fixed electrodes of iron very low in carbon, the melting point of which is higher than that of ordinary steel. This crucible is surrounded by a thick layer of asbestos, which is, in turn, imbedded in a layer of some heat-insulating material, the whole being held together by a steel case. The walls of the furnace are made so thick in relation to the dimensions of the crucible that the steel case of the apparatus may be touched with the hand without injury after having been in operation for 10 hours, at a temperature of 2370 degrees F.

The soft iron supply conductors to the electrodes are connected to the secondary copper bars of a regulating transformer which transforms the normal voltage to the low voltage (5 to 70 volts) employed in the operation of the furnace. A typical arrangement of the equipment of a large works has the furnaces provided with a hood in a central position, and a quenching tank immediately beside the furnace on one side. By this latter arrangement the change in temperature caused by carrying pieces from the furnace to the water tank is reduced to a minimum. The tank is supplied with heating and cooling coils with steam or cold water, so that the temperature of the quenching bath can be easily regulated.

A pure metallic salt or a mixture of several salts is placed in the crucible and melted by the passage of an electric current. Potassium chloride which fuses at about 1425 degrees F. is selected for carbon steel; for high-speed steels, barium chloride, which fuses at 1740 degrees F., is employed. Mixtures of these two salts will give all intermediate temperatures. For low temperatures, say between 400 and 750 degrees F., potassium and sodium nitrates may be used, and for very high temperatures, magnesium fluoride and fluorspar. The salt is melted by a movable electrode and a small piece of arc-light carbon placed in the circuit between one of the fixed electrodes and the movable one. Sparking between the carbon and the movable electrode

---
* MACHINERY, September and December, 1908.

causes salt immediately adjacent to melt and very soon a circuit is set up through a part of the salt. As the movable electrode is gradually drawn away, it leaves behind a streak of melted salt, which is extended by degrees to the opposite electrode. When this point is reached, the fusion of the remainder of the salt proceeds at a rapid rate. The temperature produced depends on the voltage employed, and may be varied by changing the intensity of the current, which is accomplished by means of a regulating transformer.

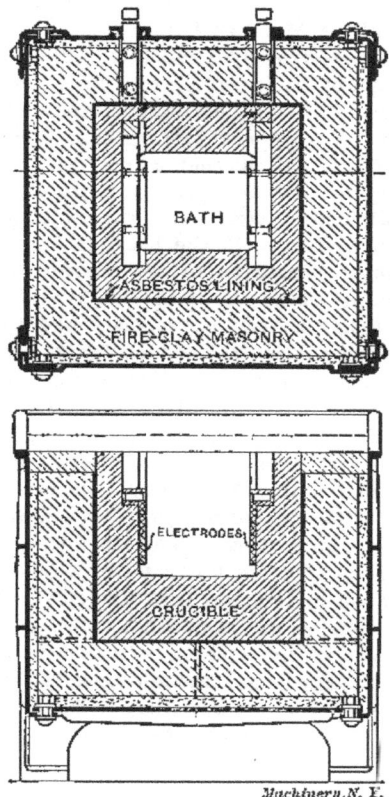

Fig. 21. Arrangement of Crucible and Electrodes for Electrically Heated Hardening Furnace

Even at a temperature of 2400 degrees F., attainable in laboratory tests but not usually employed in commercial hardening, the damage to the crucibles of the electric furnace is very small. Working ten hours a day with this temperature, a crucible will last six months, and for ordinary hardening temperatures, fifteen months.

The hardening process used by the firm of Ludwig Loewe & Co., Berlin, consists of an electrically heated barium salt bath, the arrangement of the crucible and the electrodes being as shown in Fig. 21. By means of this process, it has been possible to harden large milling cutters in about half an hour, including the time for preheating, which takes the greatest part of the time. Bringing the cutters up to a temperature of 750 degrees F. constitutes this preheating. After that, it takes only about a minute to bring an average-sized cutter to 1400 or 1500 degrees F., and then another minute to bring it up to about 2370 degrees F., which is, by this firm, considered the right hardening temperature. The time stated above refers to average-sized and heavy milling cutters, whereas it only takes from 6 to 10 minutes to bring a small milling cutter to the right temperature in the electrically heated salt bath.

The advantage of electrically heated salt baths is stated as being the total absence of any scale on the tool thus hardened, and that the tools are not distorted in the hardening process. The bright appearance is retained by the hardened tool, so that it is sometimes difficult to tell from the appearance whether a tool has been hardened or not.

In regard to cooling the cutters, the firm of Ludwig Loewe has found that when high-speed steel tools are cooled in an air blast, any moisture coming in contact with the hot tool has a tendency to crack it, so, that it becomes necessary to dry the air before it enters into the nozzles. It has also been found that it is absolutely impossible to cool a cutter which has a very heavy body and fine teeth in the air blast, as the heat from the central portion is not extracted fast enough, and therefore does not permit a sufficiently rapid cooling of the teeth to insure proper hardening. For this reason, the firm has adopted a method of cooling the cutters from the hardening heat of 2370 degrees F. to a temperature of about 1100 degrees F. by quenching in an electrically heated salt bath. After having been cooled to about 1100 degrees F. in the bath, the cutters are allowed to cool down slowly in the air, and the whole process has the advantage of being cheap and reliable, as well as effecting a considerable saving in time.

It must, however, be understood that electrically heated barium salt baths are advantageous to use only when a large quantity of tools is to be hardened, because this method will otherwise prove expensive. It has also been remarked that the electrically heated bath is more advantageous for heavy than for small tools, but it is not clear why the process should be thus limited to the former class of tools.

# CHAPTER VII

## MISCELLANEOUS HARDENING METHODS AND SUGGESTIONS

### The Gaging of Heats for Hardening*

It takes an experienced man to gage the heat for hardening with the eye, at all times, and under all conditions, without heating some tools just a little hotter than they should be to get the best results. There are various reasons why heats are not always gaged correctly. In the first place, the man has no gage to go by, and again, the light conditions, prevailing at the time, may interfere. This can be overcome by having shades at the windows, that can be adjusted. Sometimes the eye gets "off-color" and needs rest for a few minutes. The use of the "magnetic influence" in gaging the heats for hardening has therefore a great deal in its favor. This method is practically new; in fact, a great many experienced mechanics, versed in the handling of carbon steels, have never heard of it.

We all know that the proper heat for hardening steel to get the best and most lasting results, is the lowest heat at which the steel will harden. How are we to determine this? By the use of the magnetic needle, no matter what the make, or brand of steel. The gaging of heats by the magnetic needle, is done in such a way that every piece can be tested; or one may test every second, or third or fifth piece, or only the first piece and then, noting the color very carefully, harden several pieces and make no mistake. In general, the work can thus be carried on without constant reference to the magnetic needle, provided the steel is of the same carbon content, but in all cases where the carbon content changes, the test must be made again.

Mr. George T. Coles of Decatur, Ill., has experimented extensively along these lines, and describes his experiments as follows:

"In starting my experiments, I bought a small pocket compass with a jeweled pivot and needle stop, about 2 inches in diameter, and costing a dollar and a half. I got a wooden steel to rest it on close to the furnace, as shown in Fig. 22. The stool and compass should be set in such a position that the natural swing of the work back and forth when testing, will be in a plane at right angles to the needle; in other words the piece being hardened should be swung east and west. By passing the tool being hardened (in this case a milling cutter) forward and backward close to the compass, the magnetism of the tool will cause the needle to be deflected first one way, then the other, and the tool will continue to deflect the needle until the right degree of heat has been obtained, that is, the proper heat for dipping in the bath. Briefly, the right heat is reached when the tool loses its magnetism. It does not follow that if the needle remains stationary, the

---
* MACHINERY, April, 1908.

first time you test, that the heat is right, because, after the tool has reached a certain degree of heat the magnetism leaves the steel, so there is no influence on the needle, and the steel may be too hot. The different carbon contents and different grades of steel require different degrees of heat, and the magnetism leaves the steel at a certain degree of heat to correspond to the different points of carbon in the steel, and in every case this is the proper heat for dipping. After testing until the right degree of heat is obtained, I put the tool back in the furnace for about twenty seconds, just to even up the heat. I then dip the

Fig. 22. Testing a Milling Cutter, when Hardening, by Swinging it Back and Forth past a Magnetic Needle

tool in a water bath to set the hardness, and then remove it from the water bath to a lard oil bath where it remains until cold."

This method applies only to tools that can be heated all over, as it is obvious that heating a tool, say a tap, on the end to be hardened only, would have a disturbing effect on the needle because only part of the tap would be hot, leaving magnetism in the shank. In using the magnetic needle, if it points due north and south, a large body of metal would deflect the needle to a certain extent from its natural position, but no matter what the position of the needle, the moment a tool is held close to it, it is influenced by the tool, and will move until the right degree of heat is reached.

## Hardening Without Cracking*

The following method for hardening punching dies is recommended by Mr. Frank E. Shailor of Great Barrington, Mass. He has used this method of hardening successfully for more than a dozen years.

The first operation is heating. Do not prepare a die for the fire by plugging up the screw holes with fire clay, for when the die becomes warm the water is evaporated from the fire clay, causing the clay to shrink away and allowing steam and hot water to get in between the clay and hole, thus causing more trouble than if the hole had been left open for the water to flow freely through it. Place the article in the fire, with a slight draft blowing, and change its position frequently to insure uniform heating. Too much importance can not be attached to even heating, as this is the ground work of successful hardening. It should be borne in mind that steel expands nearly one-eighth inch

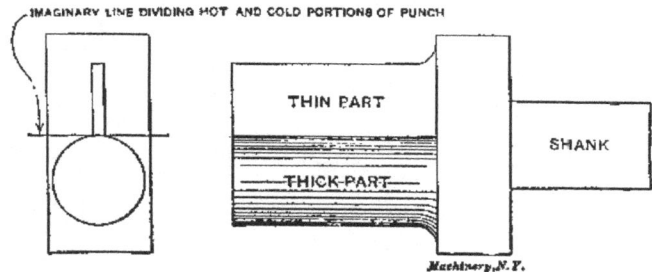

Fig. 23. Example of Work which must be Hardened with Care to Avoid Cracking

to the foot when at bright heat. Therefore, if one-half of a die is dark red and the other half is bright red, it is clear that one-half of the die has expanded more than the other half, thereby causing forces to push against each other. The condition tends to warp the die while it is hot; and if the die is dipped while in this state, as is too often done, there will be uneven contraction, causing internal strains, strongest along the line that divided the two shades of red, and causing a "set" in the die that holds it in a curved line caused by one end of the die expanding more than the other and pushing that end out of a true line.

Assume that we are to harden the punch, Fig. 23. We heat this punch carefully and evenly in a charcoal fire or gas furnace. For a bath we have a tub of clear water. The punch is removed from the fire the instant the proper heat is attained. We now immerse the punch in water, but do not allow the water to remain at the same level on the punch, as this causes either a crack or a huge swell. Now place one hand down in the water and keep feeling of the punch. The moment that the punch is not too hot for the fingers, remove it from the bath. Under no consideration allow the punch to become cold in the bath, for this is just what causes cracks, for this reason: When we immersed the punch, the slender stem was immediately attacked

* MACHINERY, February, 1906.

by the water, contracted to its normal size and became cold and hard, while the body of the punch was still red hot. This caused a distinct line between the hot and cold portions of the punch shown by the imaginary line in Fig. 23. Now, as the body of the punch begins to cool and contract, it is obvious that it must be pulling away from the stem, which has already contracted. If a crack does not appear, it is no sign that it will not sooner or later, as there must be a tremendous internal stress which will crack the punch later, and the operator will be taken to task for his carelessness. But, on the other hand, if we remove the punch as soon as we can bear our fingers on it in the

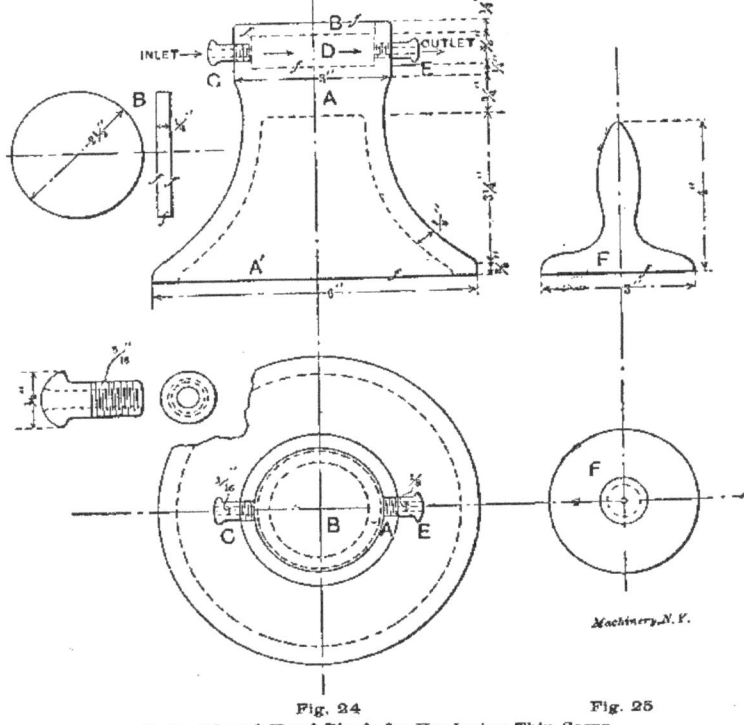

Fig. 24    Fig. 25
Pedestal and Hand Block for Hardening Thin Saws

bath (when it is still so hot that it steams), then the heat from the body of the punch will run out into the slender portion and the contraction will be uniform, excepting for the little effect the atmosphere may have.

The above could be summarized in a very few words, thus: "To prevent dies from cracking, do not allow them to become cold in the bath."

### To Harden and Temper Thin Circular Saws Without Warping*

After hardening the very thinnest saws possible to make, it is not the easiest thing to straighten them, but by employing the following

---

* S. C. Smith, MACHINERY, May, 1906.

# KINKS AND SUGGESTIONS

method there will be no trouble. It is not necessary to use oil or other preparation in hardening the saws, but simply a cast-iron pedestal, Fig. 24, through which flows a stream of water to keep it cold, and the hand block, Fig. 25.

A is the cast-iron pedestal, having its base A' reduced by coring. The top of the pedestal is bored for chamber D and receives the cast iron disk B, which is faced parallel and pressed tightly into place. Before B is pressed into place, however, holes for C and E are drilled and tapped into the casting A, these holes being made to receive brass nipples to which are attached rubber hose. Rubber hose is used inasmuch as it saves time in making the connection at the faucet and is most convenient to take down. C, being the inlet, is made larger than the outlet E, thus causing the chamber D to be constantly filled with

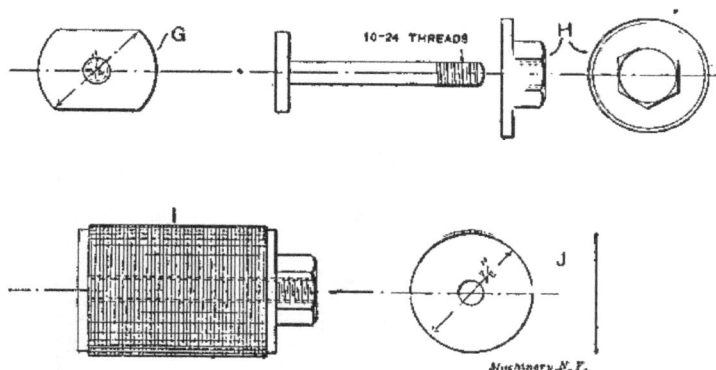

Fig. 26. Method Used for Tempering Thin Saws

water. The flow of water can be regulated at the faucet, and where there is no water faucet at the forge, a tank may be improvised by placing a pail of water above the pedestal, having a nipple connection on the side of the pail one inch above the bottom. A pail below the pedestal can serve as a drip tank. In this way a circulation of water through chamber D is assured and B is kept cold.

When sufficiently heated, the saw is laid on B and the hand block F is instantly placed on top, thus cooling and pressing it flat. The hand block should not remain on the saw more than five seconds. While another saw is being heated the operator removes the hardened saw and any scale which may have accumulated. After all the saws are hardened, the hose and fittings are "shelved" until another lot is ready.

The next operation is the tempering of the saws. In drawing the temper, use a gang arbor G, as shown in Fig. 26, which fits the hole in the saws closely. Place the saws on this arbor, as many being put on as possible, making allowance for a polished washer and nut, H. The saws are then clamped together by screwing up the nut as tightly as possible.

## No. 46—HARDENING AND TEMPERING

The saws should be drawn slowly, and should be kept revolving, now and then being dipped in water. This process is continued until the desired temper is shown by the polished washer, which should be repolished after each operation.

This method is used for very thin saws, from 0.0035 inch to 1/64 inch thick. For saws 1/64 inch thick, or thicker, it would be advisable to use oil, putting on just enough to keep the top of the pedestal well oiled.

### Hardening Shear Blades

As a rule the hardening of shear blades, such as shown in Fig. 27, is attended by more or less trouble when attempted in a shop not specially fitted for it; yet if care is exercised and a method used which is adapted to the character of the piece, the trouble from springing

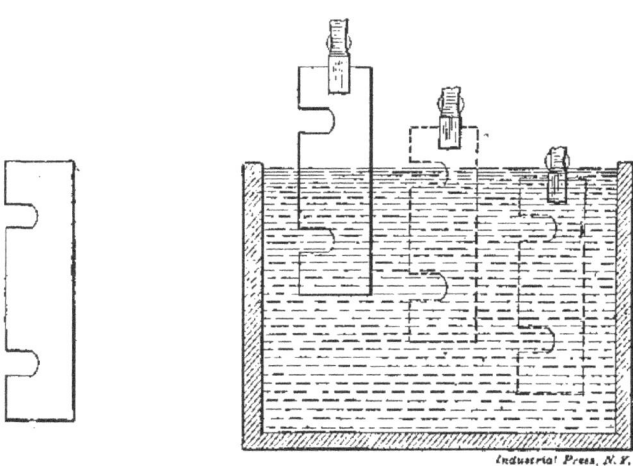

Fig. 27. Shear Blade to be Hardened

Fig. 28. Method of Dipping Shear Blade when Hardening

may be materially reduced in most cases. If the blade is made from a comparatively high carbon steel the tendency to spring will be greater than if a low carbon steel is used; but the blade will last a great deal longer if made of high carbon steel. When heating for hardening, place the blade in a hardening box and heat this in a furnace, if one is at hand. In this manner a uniform heat may be attained. If obliged to heat in a forge of the ordinary type, build a large, high fire, so as to have as nearly as possible a uniform heat. The edges and corners should be no hotter than the balance of the piece, or the uneven heat will cause the blade to crack from uneven contraction when plunged in the bath.

The bath used should be brine which is of a temperature of about 60 degrees F. Grasp the blade by one end with a suitable pair of tongs, dipping it in the bath in a vertical position, as shown in Fig. 28. It should be lowered slowly into the bath and moved edgewise while being lowered, as shown by the dotted lines. This brings the cutting edge in contact with the fresh, cool contents of the bath and allows the steam which is generated to pass from this edge to the one containing the slots. The steam thus leaves the cutting edge so that it can be acted upon by the bath, and forms a cushion at the other edge, preventing it from hardening as rapidly as it otherwise would, thus reducing the tendency to springing to the minimum.

## Composition of Quenching Baths for Tempering Cutting Tools*

The composition of a number of baths for quenching and tempering cutting tools was given by H. Le Chatelier, in an article in *Revue de Metallurgie*. Fused nitrates of potassium and of sodium are too high in temperature for certain cutting tools, as they do not permit of cooling below 220 degrees F. Mixtures of nitrate of potassium and of nitrate of sodium can, however, be employed, and a series of mixtures, fusing at different temperatures, can be obtained. He gives the following proportions for these mixtures:

| Temperature, Degrees F. | Nitrate of Potassium. | Nitrate of Sodium. |
| --- | --- | --- |
| 280 | 0 | 100 |
| 230 | 20 | 80 |
| 172 | 40 | 60 |
| 137 | 55 | 45 |
| 145 | 60 | 40 |
| 225 | 80 | 20 |
| 335 | 100 | 0 |

Higher temperatures than 400 degrees F. cannot be obtained with these mixtures. At 400 degrees F. potassium nitrate freely decomposes, whereas for steels where without extreme hardness absolute absence of brittleness is necessary, 500 degrees F. to 600 degrees F. are temperatures more suitable. The following bath gives, on fusion, a temperature of 500 degrees F.:

| | |
| --- | --- |
| Sodium chloride | 1 part |
| Potassium chloride | 1 part |
| Fused calcium chloride | 2 parts |
| Hydrated barium chloride | 1 part |
| Hydrate strontium chloride | 3 parts |

For a bath fusing at 700 degrees F., the following mixture may be used:

| | |
| --- | --- |
| Hydrated boric acid crystals | 1 part |
| Silver sand | 1½ part |
| Anhydrous potassium carbonate | 1 part |
| Anhydrous sodium carbonate | 1 part |

When prolonged treatment is required a little cyanide of charcoal may be added to prevent superficial decarbonization; but in view of the strongly cementating action of cyanide, this salt must be used with caution.

---

* MACHINERY, May, 1905.

## Temper Colors and Temperatures and Colors for Hardening*

The following tables of temper colors, and temperatures and colors for hardening, are published in a booklet issued by the Halcomb Steel Co., Syracuse, N. Y., and Chicago, Ill. The temperatures tabulated are a result of personal investigations made by Mr. Garson Myers, manager of the Chicago branch; a gas furnace equipped with a pyrometer was used. After the records were made, they were tested by two experienced tool steel hardeners, one using an electric heating furnace with a pyrometer and the other a magnetic heating furnace also connected with a pyrometer.

### Heat Temperatures and Colors for Hardening

| Degrees C. | Degrees F. | Colors. |
|---|---|---|
| 400 | 752 | Red heat, visible in the dark. |
| 474 | 885 | Red heat, visible in the twilight. |
| 525 | 975 | Red heat, visible in the daylight. |
| 581 | 1077 | Red heat, visible in the sunlight. |
| 700 | 1292 | Dark red. |
| 800 | 1472 | Dull cherry red. |
| 900 | 1652 | Cherry red. |
| 1000 | 1832 | Bright cherry red. |
| 1100 | 2012 | Orange red. |
| 1200 | 2192 | Orange yellow. |
| 1300 | 2372 | Yellow white. |
| 1400 | 2552 | White welding heat. |
| 1500 | 2732 | Brilliant white. |
| 1600 | 2912 | Dazzling white (bluish white). |

The heat and temper colors, given below, to which tools should be drawn, were contributed by a hardener and temperer of long experience, working on all grades of tool steels.

### Heats and Temper Colors of Steel

| Degrees C. | Degrees F. | Colors. |
|---|---|---|
| 215.6 | 420 | Very faint yellow. |
| 221.1 | 430 | Very pale yellow. |
| 226.7 | 440 | Light yellow. |
| 232.2 | 450 | Pale straw yellow. |
| 237.8 | 460 | Straw yellow. |
| 243.3 | 470 | Deep straw yellow. |
| 248.9 | 480 | Dark yellow. |
| 254.4 | 490 | Yellow brown. |
| 260.0 | 500 | Brown yellow. |
| 265.6 | 510 | Spotted red brown. |
| 271.1 | 520 | Brown purple. |
| 276.7 | 530 | Light purple. |
| 282.2 | 540 | Full purple. |
| 287.8 | 550 | Dark purple. |
| 293.3 | 560 | Full blue. |
| 298.9 | 570 | Dark blue. |
| 315.6 | 600 | Very dark blue. |

## To Prevent Hot Lead Sticking to Work

To prevent hot lead sticking to the work, mix common whiting or cold-water paint with wood alcohol, and paint the part that is to be

---
* MACHINERY, December, 1908.

# KINKS AND SUGGESTIONS

hardened. The hot lead will not stick, no matter how long the piece is held in the pot. Water will do as well as alcohol for mixing with the paint, but alcohol is the most convenient, inasmuch as it can be used without waiting for the paint to dry. If water is used, the paint must be thoroughly dry, as otherwise the moisture will cause the lead to fly.

## Hardening Drop Forging Dies*

On the subject of hardening drop forging dies, Mr. J. F. Sallows has contributed the following to MACHINERY:

Uneven heating and uneven cooling, with consequent uneven contraction, is the cause of so many drop forging dies cracking in hardening. There is no necessity for this trouble if the dies are properly handled. If drop forging dies are made from machine steel, they should be

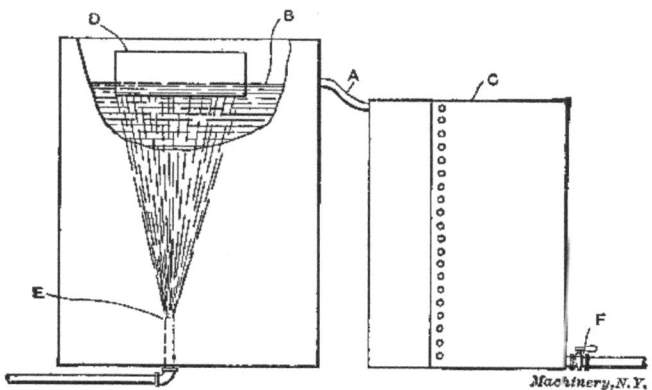

Fig. 29. Arrangement of Brine Tank for Hardening

packed in No. 1 raw bone and fine wood charcoal, three parts charcoal being used for each two parts raw bone. They are then heated in an oven for eight hours, at a temperature of 1600 degrees F., and are then dipped the same as described in the following for tool steel. When the dies are made of tool steel, the heating of the dies in an open furnace, even if covered with coke, is very injurious to the steel, as the carbon is removed from the surface of the steel, and the dies will not harden on the outside, but will be harder further in. This does not matter so much with tools that are to be ground to size after hardening, but it is poor practice with any kind of tool steel tools. Tool steel dies should be packed in fine wood charcoal in a box large enough to allow plenty of charcoal between the die and the box walls, say about two inches or more. Seal the cover on tight with asbestos cement, place the box containing the die in the furnace, and, if a pyrometer is attached to the furnace, hold the furnace at about 1500 degrees F., leaving the die in for at least four hours. For a small die, shorter time will be sufficient, but a die weighing 50 pounds or more should be allowed four hours to heat slowly and uniformly. Then,

---
* MACHINERY, January, 1908.

instead of immersing the whole die in a tank of cold, clear water, have two tanks, a large one and a small one, as shown n Fig. 29. An overflow pipe or hose $A$ from the water line $B$ in the large tank should connect it with the small tank $C$. When ready to dip the die $D$, place the face only in the water. Plenty of salt should be well dissolved in the water, about 4 pounds to the gallon; this extracts the heat from

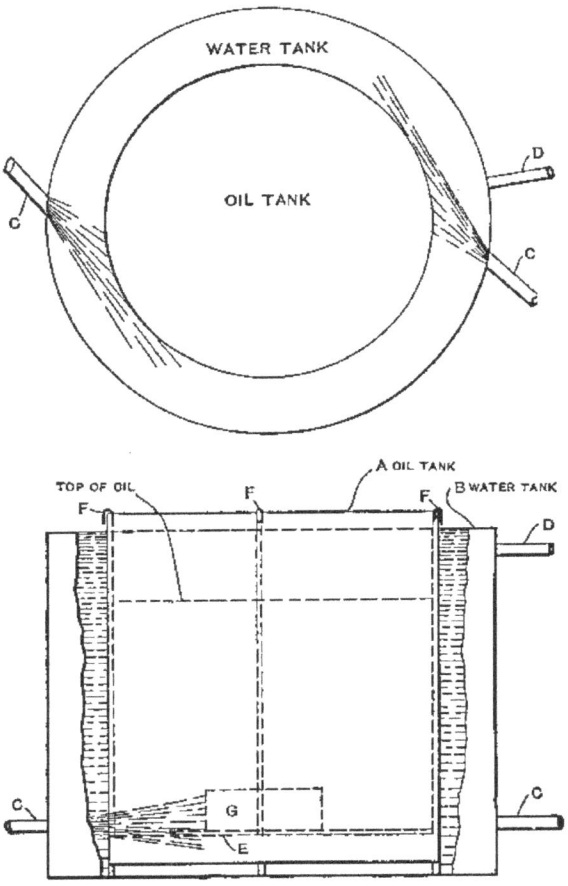

Fig. 30. Oil Tank for Hardening Room

the die quicker than clear water, and prevents steam formation on the face of the die. A water pipe $E$ should be carried in at the center of the large tank at the bottom, and should be supplied with water at fairly high pressure. When placing the die in the bath, open the valve of the pipe $E$, thus forcing the cold solution against the face of the die, while the warm water passes into the smaller tank. The solution collecting in the smaller tank, when cool enough, can be used for smaller tools, and, when so desired, can be run off by outlet $F$. An-

## KINKS AND SUGGESTIONS

other bath, in an oil tank, inside of a water tank, as shown in Fig. 30, should be provided. The size of the tanks must be determined by the size of the dies to be hardened. Fish oil should be used in this latter tank, and the tank should have two water inlets $C$, at opposite sides of the tank, and so arranged as to allow water to flow around all sides of the oil tank as indicated in the plan view. Pipe $D$ is the overthrow. A coarse mesh sieve $E$ is suspended in the oil tank, and held by rods $F$. The oil tank should have four legs about 6 inches long, to allow water underneath the tank. When the die face has been cooled in the salt water solution, remove the die quickly to the oil tank, and lower it until it rests on the sieve (see $G$, Fig. 30). Let the die remain in this position until cold. Dies hardened in this manner will not crack.

### To Anneal Spots in Hardened Saws

An easy way to anneal spots in hardened saws is to take two pieces of, say, 1-inch rod machine steel about 2 feet long and bring the ends thereof up to a white heat. Then, having previously laid the saws on a flat surface of some sort and marked the spots to be annealed with a bit of red lead, hold the heated end onto the spot a fraction of a minute. While one rod is being used the other is being heated.

### Annealing High-speed Steel*

A cast-iron box, large enough to permit proper packing of the pieces to be annealed is used. Charcoal ashes or cast-iron chips may be used for packing. Pack the work carefully, placing the larger pieces to the outside of the box, and the smaller pieces in the center. After the pieces are packed, they are then ready for the furnace. Heat slowly, raising the temperature to 1470 degrees F. (dull cherry red). Then hold the heat at this point for about 5 hours, and finally raise the heat in the furnace to 1650 degrees F. (cherry red). Shut off the fire, close the door, and let the furnace cool for 12 hours. The entire heating can be done in 5 hours, and the steel can be worked as nicely as any annealed by the steel mills. This is not the only method of annealing, but it is the best method when the steel is considered.

Annealing after hardening of high-speed steel can be accomplished by the following method in about one hour. Where a change in the tool is required to be done quickly, I often take the tool and heat to 1290 degrees F., then let it cool in the open air. Then heat the tool again, raising the temperature to 1290 degrees F. (somber red), and hold it there for 40 minutes. It is then taken from the fire, and permitted to cool in the open air. When one has at one's disposal 5 hours in which to anneal, however, the heat anneal is preferable. This is done by heating the tool or piece constantly for 5 hours. After the piece has been heated for 5 hours at 1290 degrees F. or less, take it from the fire and let it cool in the open air. One can also raise the heat to 1470 degrees F. (dull cherry red) and put the tool in lime to cool. Do not raise the heat to this degree, however, unless the piece is to be placed in lime. These methods are used only where a loss of

---

* C. U. Scott, MACHINERY, November, 1907.

time is to be considered. It is possible to anneal high-speed steel at as low a heat as 977 degrees F., red visible in daylight, but this heat will not make the steel very soft.

There is also another method of annealing high-speed steel. That is where a lead bath is in use. Take the piece to be annealed and place

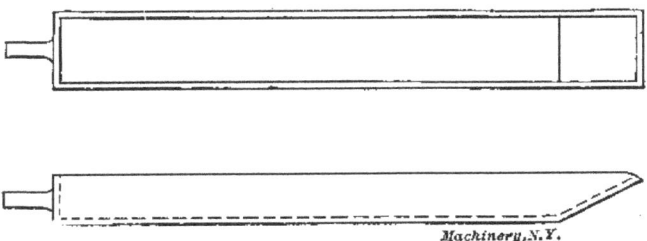

Fig. 31. Cast-Iron Box in which Blades are Packed and Heated

it in the lead while it is at a dull cherry red. If the lead is of sufficient bulk to hold the heat for several hours, it will not be necessary to continue the heat. Leave the work in the lead and remove it the next morning, by heating the lead again. Remove the tool as soon as the lead is hot enough not to adhere to the work. Then dip the work in oil, after which, returning to the lead bath, the lead will leave the

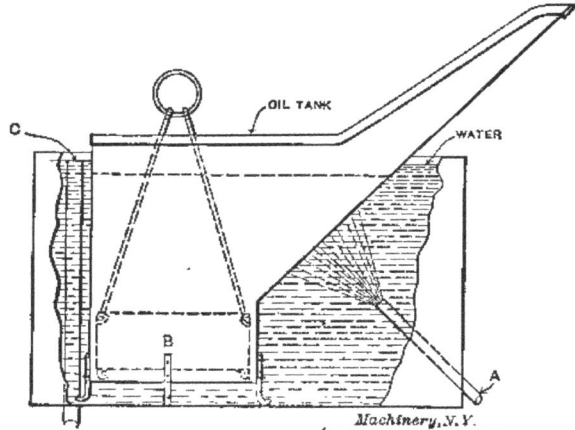

Fig. 32. Oil Tank for Hardening Hack-saw Blades

work if it is removed just as the oil burns off. After this the piece should be allowed to cool in the open air.

### Hardening Hack-saw Blades in Quantities*

Hack-saw blades may be hardened in large quantities by using cast-iron boxes of the style shown in Fig. 31 in which to heat them. The boxes should be large enough to accommodate about three dozen blades placed on edge with the back down. A little charcoal should be used at the sides to keep the teeth of the outside blades from coming in contact with the sides of the box. The blades are then placed in the

---

* James Cran, MACHINERY, October, 1908.

## KINKS AND SUGGESTIONS 47

muffle of a furnace and allowed to remain until they have reached the proper temperature for hardening. They can then be removed with a pair of tongs made to fit the shanks on the ends of the boxes. The blades should then be carefully dumped on the inclined chute of a linseed oil bath which is shown in Fig. 32. The tank containing the oil is placed inside a wooden tank that is filled with water which keeps the oil from getting overheated. Water is supplied through pipe $A$, and strikes directly on the lower side of the chute down which the blades slide on their way to the bottom, where they collect in a wire or perforated sheet metal basket $B$. An overflow pipe $C$ is placed at one

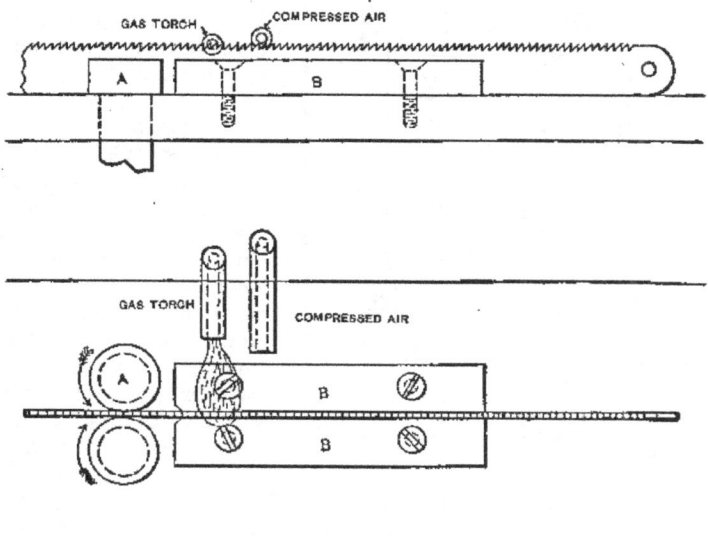

Fig. 33. Method of Hardening Flexible Hack-saw Blades

end of the water tank to carry off the warm water which rises to the top. The oil tank should rest upon legs several inches long, so as to raise it above the bottom of the water tank to allow a free circulation of the water. When the blades are fairly cooled off, the basket containing them can be removed from the oil and allowed to drip over the tank until most of the oil has left the blades; they can then be thoroughly cleaned by being immersed in a soda kettle or by placing them in clean sawdust. Flexible blades, being hardened on the teeth only, are treated differently. A fixture of the style shown in Fig. 33 is used for this method of hardening. The blades are placed, back down, between two power-driven rolls $A$ which rotate in different directions, and which feed the blades, by friction, between two guides $B$ and past the flame from a gas torch which heats the teeth sufficiently for hardening. A compressed air jet strikes the hot teeth immediately after they pass the torch. The temper does not have to be drawn, except at the ends, which is usually done with a torch.

## Apparatus for Hardening Milling Cutters

Fig. 34 shows an apparatus for hardening milling cutters, patented by Mr. J. M. Gledhill of Armstrong, Whitworth & Co., Ltd., Manchester, England. As is well known, the best results in hardening milling cutters are obtained when only the teeth are hardened and the interior of cutter is left comparatively soft. Instead of plunging the cutter directly into the water or other cooling liquid, the exterior of the cutter only is subjected to the action of the cooling liquid in the form of a spray through a series of fine jets at high pressure. The interior is kept soft by previously plugging the hole to protect it from the sudden chilling action. The cutter is made to revolve during the hardening or sprinkling process in order to expose thoroughly every part of the teeth to the spray. The chilling action may be stopped at any desired point by turning off the water, and the cutter is then gradually cooled by immersion in oil or other suitable cooling medium. This method also almost entirely eliminates the cracking of cutters in hardening.

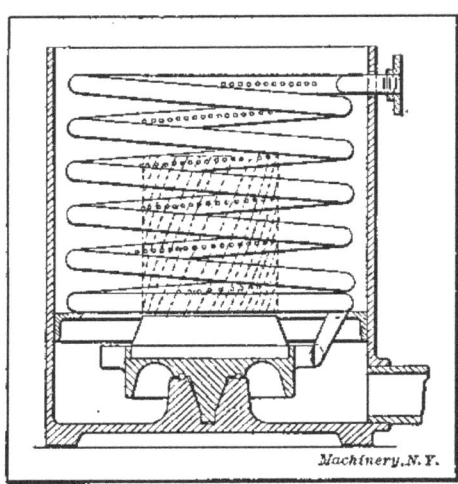

Fig. 34. Hardening Milling Cutter Teeth by a Spray of Cooling Liquid, leaving the Central Portion comparatively Soft.

The illustration clearly indicates the construction of the apparatus with a coiled pipe surrounding the cutter which is shown in dotted lines. The cutter rests on a carrier mounted on a ball bearing. The carrier is provided with vanes so that it may be caused to revolve by the action of a jet of water from the end of the coil.

## Hardening High-speed Steel[*]

It is a well-known fact that high-speed steel made up into small tools such as taps, cutters, etc., must not be blistered or pitted, as this would spoil the cutting edges. One of the principal precautions necessary, in hardening high-speed steel, to prevent blistering, is to keep it from the air while the steel is hot. A powder which is composed of corn meal, salt, and prussiate of potash mixed in equal parts has been found a very satisfactory composition. To harden the piece, heat it to a dull red and lay it in the powder. Then return the piece to the fire and heat to a bright red. Dip again in the powder and return once more to the fire and heat to almost a yellow color. Then quench the piece or pieces in cotton-seed or linseed oil.

---

[*] W. C. Betz, MACHINERY, August, 1910.

Made in United States
Orlando, FL
26 November 2023